X ARF-0571
0209

TAKE·HOME science

TAKE·HOME science

Independent Activities for Science and Technology

JENNY FEELY

Heinemann
Portsmouth NH

Heinemann
A Division of Reed Elsevier Inc.
361 Hanover Street
Portsmouth, NH 03801-3912

Offices and agents throughout the world

Copyright © 1994 Jenny Feely

The material in this book is copyright. The blackline master sheets
may be copied for classroom use within the educational institution
to which the purchaser is attached without incurring any fee. No other
form of copying is allowed without written permission. Copyright owners
may take legal action against a person who infringes their copyright through
unauthorised copying. Inquiries should be directed to the publisher.

The acknowledgements constitute an extension of this copyright notice.

ISBN 0-435-08365-1
Published simultaneously in the United States
in 1994 by Heinemann
and in Australia by
Eleanor Curtain Publishing
906 Malvern Road
Armadale, Australia 3143

Production by Sylvana Scannapiego, Island Graphics
Edited by Living Proof—Book Editing (Deborah Doyle)
Design by Patricia Tsiatsias
Illustrations by Stephen Feely, Julian Bruère
Cover design by David Constable
Cover photographs by Sara Curtain
Typeset in 13/16 pt Garamond 3
Printed in Australia by Impact Printing Pty Ltd

Contents

Part One: About take-home science

Introduction	3
Take-home science—what is it?	4
Investigation format	6
Recording and evaluating: using the log book	7
Organising the program as a take-home-science program	10
Organising the program as a classroom program	11
The resources necessary in order to conduct the program for one class group	12
The key ideas developed in the investigations	14

Part Two: The investigations

Investigations 1 to 5: Key ideas, and resources needed	21
Area of focus: Structures	
Investigation 1: Strong towers	22
Investigation 2: The newspaper cubby house	23
Investigation 3: Straw structures	24
Investigation 4: Folding furniture	25
Investigation 5: The balancing act	26
Investigations 6 to 10: Key ideas, and resources needed	27
Area of focus: Packaging	
Investigation 6: Strong packages	28
Investigation 7: Keeping water in	29
Investigation 8: Keeping food hot and cold	30
Investigation 9: Keeping food fresh	31
Investigation 10: Space-saving packages	32
Investigations 11 to 15: Key ideas, and resources needed	33
Area of focus: Simple machines	
Investigation 11: Using wedges	34
Investigation 12: Using levers	35
Investigation 13: Rollers	36
Investigation 14: Wheels	37
Investigation 15: Inventing tools	38
Investigations 16 to 20: Key ideas, and resources needed	39
Area of focus: The human body	
Investigation 16: Human skeletons	40
Investigation 17: Lung capacity	41
Investigation 18: My beating heart	42
Investigation 19: Reactions	43
Investigation 20: Getting the picture	44

Investigations 21 to 25: Key ideas, and resources needed	45
Area of focus: The sky	
Investigation 21: The moon	46
Investigation 22: The stars	47
Investigation 23: The sun	48
Investigation 24: Clouds	49
Investigation 25: Day and night	50
Investigations 26 to 30: Key ideas, and resources needed	51
Area of focus: Playground equipment and toys	
Investigation 26: Swings	52
Investigation 27: Sliding down	53
Investigation 28: Bouncing balls	54
Investigation 29: Balls and surfaces	55
Investigation 30: Rolling down hills	56
Investigations 31 to 35: Key ideas, and resources needed	57
Area of focus: Backyards	
Investigation 31: The living things in your backyard	58
Investigation 32: Leaves	59
Investigation 33: Snails on the move	60
Investigation 34: Animals in the backyard	61
Investigation 35: Soils are soils	62
Investigations 36 to 40: Key ideas, and resources needed	63
Area of focus: Plants	
Investigation 36: Growing beans	64
Investigation 37: Growing potatoes	65
Investigation 38: Growing plants from cuttings	66
Investigation 39: Good growing conditions	67
Investigation 40: Plant parts	68

Appendix: Master sheets

Master Sheet A: Log-book record	71
Master Sheet B: Possible ways to record your findings in your log book	72
Master Sheet C: 'Activities I have completed'	73
Master Sheet D: The constellations of the northern sky	74
Master Sheet E: The human skeleton	75
Master Sheet F: Cloud formations	76
Master Sheet G: The parts of a flower	77

Index   79

PART ONE

About take-home science

Introduction

Science and technology are curriculum areas that provide students with much knowledge and many skills so crucial to their being equipped for life in our future society. These two areas stand beside English-language development and mathematics as the most important study areas provided in schools.

Unfortunately, however, most often they are the areas that are neglected, left until last or left out altogether.

Many school communities are at present striving to come to terms with the challenge of providing younger students with excellent science and technology programs.

Research clearly reveals how necessary it is for children to be actively involved in their learning—for them to be doing and finding out, not simply listening. The term 'hands on' is often used to describe this teaching approach and is synonymous with effective science programs that promote excellent learning outcomes for all students.

Research also demonstrates that recognising and building strong links between home and school is crucial to children's learning outcomes being maximised. Many schools have responded to this challenge by developing home-reading programs in which parents, teachers and children work together in a close partnership to promote reading as an enjoyable and relevant part of everyday life. The reading outcomes of the children involved in these programs have been enhanced.

Take-Home Science draws together all these considerations by developing a collection of investigations and experiments children can work on at home or at school, with a friend or with a parent, in order to develop skills and knowledge in the exciting areas of science and technology. All the investigations are open-ended and involve children doing experiments and undertaking investigations themselves. The investigations use only simple resources—materials and items of equipment—that can easily be located at home and at school so children are encouraged to view technology and science as part of the everyday world in which they live.

Take-home science—what is it?

Take-Home Science comprises a series of independent investigations suitable for children seven or more years of age. It provides suggested procedures for implementing the total program within individual classrooms or as a take-home program. It also supports teachers by listing information about the key ideas developed in each investigation and the resources that have to be stocked for the class to use in each activity. Each of the 40 activity sheets outlines the purpose of the investigation, what is needed in order to undertake it, step-by-step instructions, and open-ended but focused questions.

The program is very versatile and can be used in a variety of ways, listed as follows.
- As a take-home program
- As a classroom science and technology program
- As a classroom activity centre for fast finishers
- As a program within other activity programs
- As a program intended to promote the extension of more able students
- As a resource for cross-age tutoring
- As a support for integrated-curriculum units of study
- As part of a library borrowing system

It can be used by one classroom teacher alone or by a group of teachers.

What do the children do?

Whichever way you choose to use the program, the process for involving the children is easily understood and implemented. Children are involved in completing the following five steps, either totally within the classroom or partially at school and partially at home with help from their parents.

Step 1

Either individually or with a partner, choose an investigation to undertake and check it off on the tally sheet.

Step 2

Check all the necessary resources are available: you will have to either collect them from school or bring them from home.

Step 3

Read the investigation card (the activity sheet photocopied) and undertake the investigation, keeping any appropriate records in a learning log book.

Step 4

Report your findings either during a class 'sharing' time—by presenting a project—or by being interviewed by your teacher or one of your peers.

Step 5

Choose another investigation.

The investigations

In order to facilitate organisation, you can photocopy each activity sheet on to stiff paper or card. To protect the cards, cover them with 'Contact' or a similar clear adhesive material, laminate them, or store them in plastic pockets. If they are to be taken home, store each one in a plastic bag along with any necessary equipment. Alternatively, set up a pocket inside children's log books to which they can transfer their investigations. Children can carry the resources they need in ice-cream containers or large yoghurt containers—these are easily collected and cost nothing. They can be used to either permanently house resources for a specific investigation or to fill up with the necessary resources each time a new investigation is undertaken.

The investigations are organised into *eight* 'areas of focus' as listed in the following table.

The eight areas of focus	
Investigation numbers	**Area of focus**
• 1, 2, 3, 4, 5	Structures
• 6, 7, 8, 9, 10	Packaging
• 11, 12, 13, 14, 15	Simple machines
• 16, 17, 18, 19, 20	The human body
• 21, 22, 23, 24, 25	The sky
• 26, 27, 28, 29, 30	Playground equipment and toys
• 31, 32, 33, 34, 35	Backyards
• 36, 37, 38, 39, 40	Plants

Each investigation stands alone as a complete investigation. Across the *five* investigations in each area of focus, children have the opportunity to develop a wider understanding and to make links and draw general conclusions.

Each investigation follows the same format, as set out on page 6.

Investigation format

The title indicates what the investigation is about.

This is the key question the investigation is intended to answer.

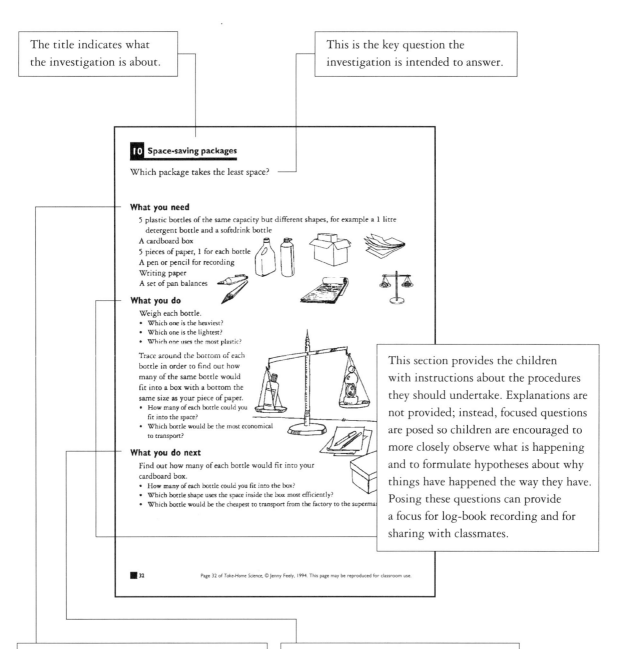

This section provides the children with instructions about the procedures they should undertake. Explanations are not provided; instead, focused questions are posed so children are encouraged to more closely observe what is happening and to formulate hypotheses about why things have happened the way they have. Posing these questions can provide a focus for log-book recording and for sharing with classmates.

This is a basic checklist indicating all the resources you will need in order to complete the investigation. The children themselves will have to collect some items, which should be readily available from home; other items will most easily be provided by the school. Ticking off the items on the checklist will ensure the child is ready to commence the investigation.

This section provides further investigations that will either take the child into related explorations or encourage him/her to use what he/she has learnt from the initial investigation in order to complete the next step. Focused questions are again provided so children are encouraged to observe and to formulate hypotheses.

Recording and evaluating: using the log book

Primary-school teachers often find it comforting to have students record their findings after completing science activities. The record is concrete evidence a child has completed the task and has been doing some 'real work', not simply playing around with balls or containers of water and so on. It may also be evidence of what the child has gained from his or her investigation. Unfortunately, it is not always an activity that children find purposeful or useful. The use of learning logs overcomes this problem because it encourages children to respond to their findings by using recording methods that are appropriate to them and that enable them to make sense of what they have learnt.

When you use learning logs the following things happen.

Children are encouraged to explore specific language.

As children are learning in any curriculum area, they have to have the opportunity to discover and use the language that is specific to the area. Children building a cardboard skeleton will be learning the anatomical names for bones such as the scapular as well as the common name, shoulder blade. They will be learning words such as 'joints', 'ligaments' and 'cartilage'. Using the learning log book provides children with an opportunity to record this specific language.

Children are encouraged to use language for their own purposes.

Children have to have many opportunities to see written language as being relevant and useful to them, to see writing as having a purpose that enables them to think and learn in ways they would not otherwise be exposed to. Using the learning log encourages children to respond to their experiments and investigations in ways that best suit their purpose.

Children are encouraged to use a wide range of language forms.

Children can respond to their experiences in countless ways. Having traditional science reports is only one of the ways and does not always best serve the learner's needs. Children whose written-English skills are less well developed may be better able to accurately record their thoughts by doing drawings. Other children may just as clearly demonstrate their understanding of a topic through writing a story or a poem.

Through exploring and becoming comfortable with a wide range of written forms, children become better equipped to choose the form that is most useful and appropriate to the investigation at hand and to their personal needs at the time. The range of responses suggested on *Master Sheet B: Possible ways to record your findings in your log book* is not exhaustive —you might wish to work through each of them with your class, by modelling the form and making explicit the construction of each type. Have the class members add to the list as they make new discoveries.

Children 'own' the learning.

Children often view the things teachers and parents ask them to do as being done to please the teacher or parent; they do not view their learning as being of direct benefit to themselves. As learners they have to be encouraged to view their actions as directly benefiting themselves. You could say to the child, for example, that 'learning to spell these words you are having trouble with will help you communicate more easily,' or 'Now you have learnt this it is easier for you to get your ideas across clearly.'

Children should not be recording their findings simply to make their teacher and parents happy, not cross; they should be recording them because the process helps them learn and is one of inherent interest to them. Questions worth considering are, 'What do you learn by doing this?', 'What can you do now that you couldn't do before?' and 'How does this learning help, change or excite you?'

Children share the learning.

Keeping records enables children to refer to and reflect on what is happening. It is a useful way of keeping track of observations, measurements, predictions and questions. Having the record enables children to compare their own results over time, to compare their results with those of other people, and to clarify their thinking. It is a vital component of science and technology investigations. Using the learning log book enables children to report to other children and to share their findings with them, accurately and in a meaningful way.

You keep track of each student's progress.

The challenge of keeping up with the learning each child is achieving while undertaking individual or paired investigations can be tackled in a number of ways, as follows.

1. Have regular sharing and reading of log-book entries. The entries can be annotated in order to identify development and can be copied for student files.
2. Before and after they have completed three or four cards from one area of focus, for example 'Plants', ask students to complete concept-maps showing what they understand.
3. Have students self-evaluate by indicating what they now know.
4. Ask students to present what they have learnt in a specific form, for example, 'Write a report/poem/play/chart showing what plants need in order to survive.'

Communication with home is promoted.

When using any take-home program it is vital that open communication between home and school be fostered and encouraged. Having a child's parents and teacher make entries in his/her log book is one way of doing this. It alerts the teacher to any problems that have arisen and enables the sharing of exciting discoveries and progress to take place. It is also offers a way of providing positive feedback to the child about his/her own efforts and learning.

Organising the program as a take-home-science program

Establishing effective take-home programs necessitates getting all the people involved together in order to share their ideas, dreams and concerns. This means that at some stage before launching the program you will have to gather together the children, their parents and any colleagues you are intending to work with. This will enable everyone involved to reach common understandings about what the program is for, how it works, and what each person's role is in it. It is useful to keep parents informed even if you do not intend to conduct the activities as a take-home program.

If you decide to use the activities as a take-home program, you might find it useful to form a team of interested teachers and parents for the purpose of amassing and organising the resources. This would spread out the workload and the enthusiasm and make success much more likely. It would also help avoid the problem of children's 'remembering' they need something five minutes before they walk out the door on their way to school.

Organising the program as a classroom program

The investigations can be used in a variety of ways within the classroom. You could set them up as an activity centre and have children work their way through the cards with a partner in a weekly science session. It is recommended that if the investigations are used in this way, children be encouraged to work through the five investigations within an area of focus before moving on to another area of focus. Alternatively, these investigations could be part of a series of cross-curriculum activities through which children rotate in groups across an English–maths–art –science activity, during an afternoon or over one whole day.

You could also present the investigations as a series of activities children work their way through in groups, over the course of several weeks. You could relate this type of investigation to a specific topic or make it part of an integrated-curriculum unit. For example, while investigating the topic 'Me', you could use investigations 16 to 20 in order to further investigate the human body. For the integrated topic 'How technology changes our world', you could use investigations 11 to 15 in order to enable children to develop knowledge about the workings of simple machines.

You could organise the investigations in one corner of the classroom as activities for fast finishers and/or interesting activities for wet days. Do this by having a box of activity cards and a box or storage area in which the necessary resources are available.

If using the program as a cross-age tutoring program, it is advisable to allow time for the older children, who will be 'teaching' the younger children, to first complete the investigation themselves. This avoids the children's becoming involved in finding out for themselves and forgetting they should be working to help other children find out. If this is the approach you take, make sure the older children clearly understand the program's purpose: they are to help by asking questions and answering any questions raised, not by doing the investigation for the younger children.

The resources necessary in order to conduct the program for one class group

The way you choose to run this program will determine which resources you need to collect and purchase in advance and which ones can be left to the children to collect. If you conduct the program as a take-home program, many of the items listed can be found at home. If you conduct it as a classroom program, you will need to consider ways to collect some materials within the classroom.

Materials for the school to buy

A supply of laminating materials for the activity cards
Wooden rulers
Pan balances
Stopwatches
Thermometers
Waterproof marking pens
Soft clay or plasticine
5 magnifying glasses
A compass
A tape measure
A ruler of length 1 metre
A bottle of nail polish
A food strainer

Materials to prepare in advance

Play dough
A cooked chicken carcass. Boil some chicken carcasses for several hours until the flesh is falling away from the bones. Separate the bones from all the fleshy parts and wash them thoroughly. Dry the bones in a slow oven or in the sun. When storing them, leave them in an open container.
A ruler with a hole drilled in one end

Expendable materials

Wax crayons
Candles
Rubber bands
Paper
Pencils
Fabric scraps
Cardboard off-cuts
Plastic straws
Bendable straws
Pipe cleaners
Re-usable adhesive (e.g. Blu-Tack)
Split pins
Thumb tacks
Masking tape
Adhesive tape
Pop-sticks
Wood glue
15 packets of bean seeds
Satay sticks (thin wooden skewers)
Polystyrene cups
Red and green cellophane paper
Yoghurt pots
Plastic wrap
Paper
String
Paper towels

Materials to collect
- Several empty film cases
- Fabric scraps
- Cardboard off-cuts
- Several balls, for example tennis balls and basketballs
- Several building blocks all the same size and shape

Materials that can be found at home
- Books
- Newspapers
- 10 ounce food tin
- A toy truck
- Several food packages, for example egg cartons and cracker packs
- A rolling pin
- Eggs
- Cooking oil
- A ceramic cup
- A tin cup
- A clock or watch with a 'second' hand
- Ice cubes
- Paper bags
- Plastic bags
- Cookies
- Several containers with lids, for example cookie tins and screw-top jars
- Several ice-cream containers
- Several plastic bottles
- Kitchen tools, for example pizza cutters and can openers
- Brooms
- Rakes
- Rolling pins
- Egg beaters
- A can opener with wheels
- A round tin
- A pair of scissors
- A flashlight
- Buckets
- Glass jars
- Potatoes with 'eyes'
- Stones
- Leaves from several plants, for example geraniums, daisies and African violets
- Soil
- Several cylindrical objects, for example tins and cups
- An alarm clock
- A shovel
- Pencils

The key ideas developed in the investigations

Investigation	Key ideas
1 Strong towers	• Triangles are stronger and more rigid than other shapes when used in building. • Tubes are strong structures that are able to support weight and resist bending and twisting.
2 The newspaper cubby house	• Triangles are stronger and more rigid shapes than other shapes when used in building. • Tubes are strong structures that are able to support weight and resist bending and twisting.
3 Straw structures	• Triangles are stronger and more rigid shapes than other shapes when used in building. • Tubes are strong structures that are able to support weight and resist bending and twisting. • The way materials are joined affects their strength.
4 Folding furniture	• Folding structures use a variety of movable joints. • Folding joints can be strengthened in various ways. • The ability to fold presents problems of stability that can be overcome. • Gravity is one force that acts on folding structures.
5 The balancing act	• Mobiles require the balancing of various masses. • For effective balance, each mass's centre of gravity should be below the support.
6 Strong packages	• Packages are designed to support and protect the objects placed inside them. • Some materials are better able than others to withstand impact. • A packet's shape affects its stability and strength.
7 Keeping water in	• Some materials absorb water and others do not. • Materials that are water resistant make good food containers. • We can make materials water resistant by coating them with oil or wax or another impermeable material.
8 Keeping food hot and cold	• Materials conduct heat at varying rates. • Some materials are good insulators. • The amount of temperature change in a substance is related to the insulating nature of the container it is in.

Investigation	Key ideas
9 Keeping food fresh	• A container's ability to keep air in or out affects the freshness of the food stored in it. • A lid's design affects its airtightness.
10 Space-saving packages	• Different-shape containers can hold the same amount of liquid. • A container's shape is in part related to the amount of material required in order to produce the shape. • Some shapes are more space efficient when being transported in cartons.
11 Using wedges	• Knives and other cutting implements are wedges. • Wedges become increasingly thick away from their edge. • A wedge is an inclined plane used in order to cut or dig.
12 Using levers	• Handles are levers. • Levers make work easier. • Levers can push or pull things. • Many household implements are designed to use leverage in order to make work easier.
13 Rollers	• Having rollers makes it easier to flatten things. • Rollers distribute the force used across a wider area.
14 Wheels	• Many household utensils use wheels. • The wheels may have gears attached. • Using wheels makes work easier.
15 Inventing tools	• People have invented many implements in order to make their lives easier. • Many tools use the same idea applied in different ways in order to achieve different outcomes. • People invent new tools in order to solve problems and difficulties as they arise.
16 Human skeletons	• Bones give structure and strength to bodies. • Bones are of different shapes and sizes. • We can see the places at which ligaments and tendons join to bones. • The ends of some bones are shaped in order to enable joints to move smoothly.
17 Lung capacity	• Muscles and the diaphragm contract and relax in order to enable lungs to fill and empty the air we breathe. • Each person's lung capacity can be measured.

Investigation	Key ideas
18 My beating heart	• Heart rate can be measured through observation of the pulse rate. • Heartbeat rate varies according to the amount of exercise being done.
19 Reflexes	• Messages sent to the central nervous system instigate responses. • With practice, the speed of a person's reaction time can be trained to increase.
20 Getting the picture	• The color of the light entering the eye determines the perception of color. • Colored filters make colors appear to be other than what they are.
21 The moon	• The moon has a regular cycle of movement around the earth. • The appearance of the moon and the time it is visible in the sky change in a regular and predictable way.
22 The stars	• We can identify and name stellar constellations. • We can see and name apparent star movements.
23 The sun	• The sun seems to rise in the east and set in the west in all countries.
24 Clouds	• Clouds have characteristics we can see and record. • We can classify clouds according to their color, shape and density.
25 Day and night	• The times of the sun's rising and setting change. • We can predict the times, and they follow a regular pattern throughout a year.
26 Swings	• Swings behave in regular and predictable ways. • A swing's height of release does not affect the number of swings in one minute. • The swing's mass does not affect its rate of swing.
27 Sliding down	• An object will take the same time to slide down the same slope each time it is released. • We can use slopes in order to move objects.
28 Bouncing balls	• A dropped ball will rebound to a height less than that it was dropped from. • As the ball strikes a hard surface it is compressed. • As the ball returns to its shape it pushes against the surface and rises.
29 Balls and surfaces	• The nature of the surface a ball is dropped on determines the height the ball will bounce. • Some materials absorb more energy than others. • The material a ball is made of will determine the height of its bounce.

Investigation	Key ideas
30 Rolling down hills	- We can use slope in order to move objects. - As a slope's gradient increases, the distance travelled by a ball rolling down the slope also increases.
31 Living things in your backyard	- We can classify all things as living, once living or non-living. - Living things have a relationship to their habitats. - We can see and record characteristics of all things.
32 Leaves	- Many plants have leaves, and leaves have a diverse range of shapes, colors and textures.
33 Snails on the move	- A snail's habitat provides the snail with shelter, moisture and food. - Snails often shelter in the same place each day. - We can record the movement of snails.
34 Animals in the backyard	- A wide range of animals live in the backyard habitat. - The number of each animal living in a habit depends on the richness of the food source, availability of shelter, climate, and presence of predators.
35 Soils are soils	- Soil is made up of a variety of organic and non-organic material. - Soil's composition changes as we dig further below the surface. - Soil is made up of small and large pieces.
36 Growing beans	- Seeds contain enough energy to start growing. - Bean seeds send out roots that move downwards and shoots that move upwards.
37 Growing potatoes	- Potatoes are tubers. - Potatoes contain enough nutrients to make a new plant start growing. - Potatoes reproduce by forming new tubers.
38 Growing plants from cuttings	- Some plants can be grown from cuttings in water. - Some plants will form roots when parts of them are exposed to water. - We can plant the cuttings in order to grow new plants.
39 Good growing conditions	- Plants need water, nutrients and some light in order to grow in a healthy way.
40 Plant parts	- Plants have parts we can identify. - Flowers can be of various shapes and colors. - Some plants have flowers that are their reproductive parts.

PART TWO
The investigations

Investigations 1 to 5: Key ideas, and resources needed

Area of focus: **Structures**

Investigation	Key ideas	Resources needed From home	From school
1 Strong towers	• Triangles are stronger and more rigid than other shapes when used in building. • Tubes are strong structures that are able to support weight and resist bending and twisting.	• Several newspapers • Several heavy books	• A roll of adhesive tape
2 The newspaper cubby house	• Triangles are stronger and more rigid shapes than other shapes when used in building. • Tubes are strong structures that are able to support weight and resist bending and twisting.	• Newspaper	• A roll of masking tape
3 Straw structures	• Triangles are stronger and more rigid shapes than other shapes when used in building. • Tubes are strong structures that are able to support weight and resist bending and twisting. • The way materials are joined affects their strength.	• A 10 ounce food tin	• 40 straws • 40 pipe cleaners cut in half—for best results, store them in a press-seal plastic bag.
4 Folding furniture	• Folding structures use a variety of movable joints. • Folding joints can be strengthened in various ways. • The ability to fold presents problems of stability that can be overcome. • Gravity is one force that acts on folding structures.	• A toy	• Straws • Pop-sticks • Thumb tacks—keep them in film cases. • Split pins—keep them in film cases. • Fabric scraps • Wood glue—small, well-sealed plastic bottles are best.
5 The balancing act	• Mobiles require the balancing of various masses. • For effective balance, each mass's centre of gravity should be below the support.	• Waterproof marker pens	• A ruler • String—4 metres • Plasticine or soft clay • 5 satay sticks with their points removed • Cardboard—off-cuts are okay.

1 Strong towers

What shape is the best for building towers?

What you need

Newspaper
Adhesive tape
Several heavy books

What you do

Make triangles and squares from sheets of rolled-up newspaper—one sheet of paper per roll.
Join the triangles together to make a tower.
Join the squares together to make a tower.
Put one or more heavy books on top.

- Which shape is best for building a strong tower?
- Which tower used the most newspaper?

What you do next

Build a newspaper tower you can stand up in that will support two books on top.

- How did you overcome any problems you had?

Look around you.
Look inside your garage.

- What shapes do you see in the garage's roof and walls?
- Why do you think these shapes have been used?

Look at buildings that are being built.

- What shapes do they use?

Look at any bridges there are in your neighborhood.

- What shapes have been used in building them?

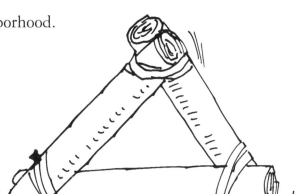

2 The newspaper cubby house

What makes a good cubby house?

What you need

Newspaper
Masking tape

What you do

Using only the masking tape and newspaper, build a cubby house you can fit inside.

- What shapes have you used in order to make your cubby house strong?
- What shapes have you used in order to make your cubby house big enough for four people?

What you do next

Build a cubby house one person can stand up in.
- What shapes did you use this time?
- How did you have to change your original cubby house?

3 Straw structures

How tall can a straw tower be?

What you need

20 plastic straws
Pipe cleaners cut in half
A 10 ounce food tin

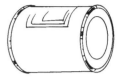

What you do

Build the tallest tower possible from 20 straws.
- How tall is your tower?
- What happens to it after one hour?
- What happens to it after one day?
- What have you done in order to make your tower strong?
- What have you done in order to make your tower tall?

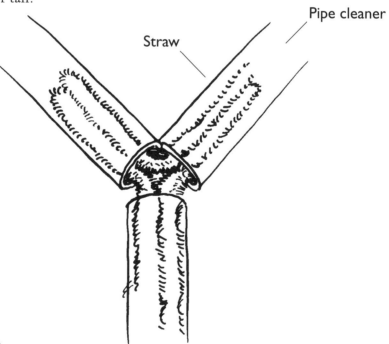

What you do next

Build a tower that can support the can of food at least 10 centimetres above the surface the straw structure is resting on.
- What have you done in order to make your tower strong?
- What have you done in order to make your tower tall?

4 Folding furniture

What is the best way to build fold-up furniture?

What you need

Plastic straws
Pop-sticks
Thumb tacks
Split pins
Fabric scraps
Wood glue
A toy, for example a toy truck

What you do

Examine any collapsible furniture you can think of.

Draw pictures of these things showing how they fold up.
- How do fold-up things work?
- What type of joints do they have?
- Which of their parts are fixed?
- Which of their parts can move?
- Which of their parts can be attached in order to give stability?
- How are the fold-up things made safe and secure when they are up?

What you do next

Build a piece of fold-up furniture for your toy.

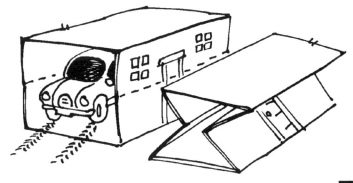

5 The balancing act

How do mobiles balance?

What you need

A ruler
String
Plasticine or soft clay
Satay sticks with their points removed
Pieces of cardboard
Waterproof marker pens

What you do

Tie the string 10 centimetres from the end of the ruler.
Use all your plasticine in order to balance the ruler so it hangs horizontally.
- How did you balance the ruler?
- Where do you have to use the most plasticine?

Try this with the string tied in a different place on the ruler.
- What happens as the string is moved closer to the centre of the ruler?
- Where do you put the plasticine, and how much do you use?

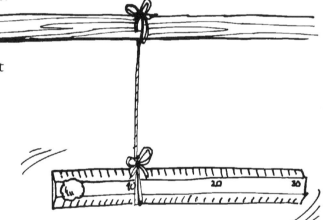

What you do next

Use this information in order to create a mobile that has at least six figures hanging.
- How do hanging objects' sizes affect the mobile's balance?
- How can you overcome imbalance when using plasticine?

Investigations 6 to 10: Key ideas, and resources needed

Area of focus: Packaging

Investigation	Key ideas	Resources needed – From home	Resources needed – From school
6 Strong packages	• Packages are designed to support and protect the objects placed inside them. • Some materials are better able than others to withstand impact. • A packet's shape affects its stability and strength.	• Containers used for packaging food, for example a milk carton, an egg carton, a cereal box and a cracker pack • A boiled egg • Cardboard off-cuts • Newspaper	• A roll of adhesive tape • Play dough—seal it in a container or a press-seal sandwich bag.
7 Keeping water in	• Some materials absorb water and others do not. • Materials that are water resistant make good food containers. • We can make materials water resistant by coating them with oil or wax or another impermeable material.	• Empty cardboard food containers, for example a milk carton, an egg carton, a cereal box and a cracker pack • Tap water • Cooking oil	
8 Keeping food hot and cold	• Materials conduct heat at varying rates. • Some materials are good insulators. • The amount of temperature change in a substance is related to the insulating nature of the container it is in.	• A ceramic cup • A tin cup *Note*: The two cups should be as close to the same size as possible. • A clock or watch • Ice cubes	• A thermometer—surround it with polystyrene to make sure transportation is safe. • A polystyrene cup
9 Keeping food fresh	• A container's ability to keep air in or out affects the freshness of the food stored in it. • A lid's design affects its airtightness.	• A paper bag • A plastic bag • Rubber bands • Crackers • 3 containers with lids, for example a cookie tin, an ice-cream container and a screw-top jar	
10 Space-saving packages	• Different-shape containers can hold the same amount of liquid. • A container's shape is in part related to the amount of material required in order to produce the shape. • Some shapes are more space efficient when being transported in cartons.	• 5 plastic bottles of the same capacity but different shapes • A cardboard box • 5 pieces of paper—one for each bottle you have • Pens and pencils for recording • Writing paper • A set of pan balances—children may have to borrow this from school.	

6 Strong packages

Which packaging is the strongest?

What you need

Food-packaging containers, for example an egg carton, a milk carton, a cereal box and a cracker pack
Play dough
A boiled egg
Adhesive tape
Pieces of cardboard
Newspaper

What you do

Try the following with several types of food container.

Carefully take the food out of the package.
Inside the package, place a ball of play dough shaped to look like the food from the package, then reseal the package.
Drop the package from a height level with your nose.
Open the package.
- What has happened to the play dough?
- How effective was the package at protecting the play dough?
- What was the package made of?
- What shape was the package?
- Did the package stop the play dough moving inside it?
- How did these things affect the play dough when it was dropped?

What you do next

Build a container that will protect an egg when it is dropped on a hard surface from a height level with your nose.
Test the container by putting the play dough in it.
When you think the experiment will work, try it with the boiled egg—or an uncooked egg, if you dare!

7 Keeping water in

Which food containers keep water in, and how do they do it?

What you need

Food-packaging containers, for example a milk carton, an egg carton, a cereal box and a cracker pack
Tap water
Cooking oil
A wax crayon
A candle

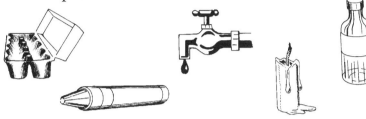

What you do

Pour 1 cup of cold water into each container.
Watch the containers for 10 minutes.
- What happens to each container?

Watch for 10 more minutes.
- What happens now?
- Does the water stay in?
- Does the container stay the same?

Look at the containers.
- How are they the same?
- How are they different?
- Why do some containers keep water in and others not keep it in?

What you do next

Take a dry cardboard container that would not hold water.
Find as many ways as you can to make it hold water.

Investigate ways in which people have made objects waterproof in the past.

Page 29 of *Take-Home Science*, © Jenny Feely, 1994. This page may be reproduced for classroom use.

8 Keeping food hot and cold

How do containers keep food hot?

What you need

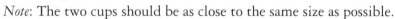

- (Hot) tap water
- A polystyrene cup
- A ceramic cup
- A tin cup

Note: The two cups should be as close to the same size as possible.

- A thermometer
- A clock or watch with a 'second' hand
- Ice cubes

What you do

Put half a cup of hot tap water into each cup.

Measure each cup's water temperature using the thermometer and by touching the outside of the cup with your hand every 5 minutes for half an hour.

- What happens to the temperature of the water in each cup as time passes?
- Which cup enables the heat to be lost most quickly?
- Which is the best insulator?
- Which was the safest cup to handle?

What you do next

Put an ice cube into the 3 empty cups.

Find out how long the ice takes to melt in each cup.

- How does the material a container is made of affect the container's ability to keep things warm or cool?

9 Keeping food fresh

What does 'airtight' mean?

What you need

A paper bag
A plastic bag
Rubber bands
Crackers
3 containers with lids, for example a cookie tin,
   an ice-cream container and a screw-top jar

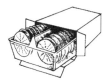

What you do

Put 1 cracker into each of the 2 bags
and close the bags tightly, using the rubber bands.
Leave the bags for 2 nights.
On the next day, eat each cracker.
- Which cracker was the freshest?
- Which bag is the best for storing food?

What you do next

Put 1 cracker into each of the 3 containers, close the containers' lids and leave them for 1 week.
At the end of the week, eat the cracker.
- Which cracker was the freshest?
- Which container is the best for storing food?
- Which container's lid was the most airtight?

10 Space-saving packages

Which package takes up the least space?

What you need

5 plastic bottles of the same capacity but different shapes, for example a 1 litre detergent bottle and a softdrink bottle
A cardboard box
5 pieces of paper, 1 for each bottle
A pen or pencil for recording
Writing paper
A set of pan balances

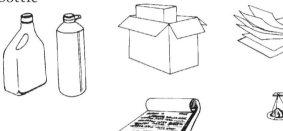

What you do

Weigh each bottle.
- Which one is the heaviest?
- Which one is the lightest?
- Which one uses the most plastic?

Trace around the bottom of each bottle in order to find out how many of the same bottle would fit into a box with a bottom the same size as your piece of paper.
- How many of each bottle could you fit into the space?
- Which bottle would be the most economical to transport?

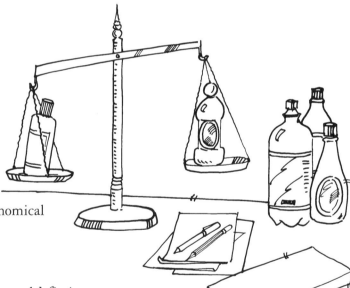

What you do next

Find out how many of each bottle would fit into your cardboard box.
- How many of each bottle could you fit into the box?
- Which bottle shape uses the space inside the box most efficiently?
- Which bottle would be the cheapest to transport from the factory to the supermarket?

Investigations 11 to 15: Key ideas, and resources needed

Area of focus: Simple machines

Investigation	Key ideas	Resources needed From home	From school
11 Using wedges	• Knives and other cutting implements are wedges. • Wedges become increasingly thick away from their edge. • A wedge is an inclined plane used in order to cut or dig.	• Plasticine or soft clay • Cardboard • Kitchen tools, for example knives, cookie cutters, pizza cutters and scissors *Safety note*: An adult has to know what the child is doing in this investigation, and may wish to supervise him/her. Take care with all cutting implements. You may wish to highlight the warning on the activity card in order to draw the child's attention to it.	
12 Using levers	• Handles are levers. • Levers make work easier. • Levers can push or pull things. • Many household implements are designed to use leverage in order to make work easier.	• A broom • A rake (or shovel)	
13 Rollers	• Having rollers makes it easier to flatten things. • Rollers distribute the force used across a wider area.	• A pencil or pen for recording • A rolling pin • Several cylindrical objects	• Plasticine or soft clay
14 Wheels	• Many household utensils use wheels. • The wheels may have gears attached. • Using wheels makes work easier.	• An egg beater • A can opener with wheels • A pizza cutter • Pieces of cardboard • A pair of scissors • A round tin • Glue	
15 Inventing tools	• People have invented many implements in order to make their lives easier. • Many tools use the same idea applied in different ways in order to achieve different outcomes. • People invent new tools in order to solve problems and difficulties as they arise.	• Newspaper • Cardboard • Adhesive tape • A pair of scissors • A pen or pencil for drawing • Paper for drawing and writing	

11 Using wedges

How do cutting tools work?

What you need

 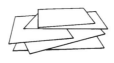

Plasticine or soft clay
Pieces of cardboard
Kitchen tools such as knives, cookie cutters, pizza cutters and scissors
Warning: Make sure an adult knows what you are doing and supervises you in this activity. Take care with all cutting implements.

What you do

Knead the plasticine until it is soft, then flatten it out.
Use the tools you have collected in order to cut the plasticine. Carefully examine the tools and the marks they have made.

- Is the tool the same width at the back of the blade as at the cutting edge?
- How does this help you cut?

What you do next

Make wedges from the cardboard.
Use these wedges in order to cut the plasticine.
- How does the wedge's thickness affect its ability to cut through the plasticine?

Compare the thicknesses of the cutting implements you have been using.
- Which implement is the sharpest?
- Which implement has the thinnest wedge at the edge?

12 Using levers

How do brooms and rakes work?

What you need

A broom
A rake (or shovel)

What you do

In your yard, choose an area that requires sweeping.
Time how long it takes you to sweep half this area holding the handle near the top.
Time how long it takes you holding the brush end of the broom holding the handle near the bottom.

- Which way was the quickest?
- Which way made you the most tired?
- How important is the broom handle in getting the sweeping work done?

What you do next

Try the same thing using a rake or a shovel.
- How does the handle help you get the job done?

Look at other tools in the kitchen or garage.
- Which of the tools have long handles in order to help make work easier?

13 Rollers

How do rollers work?

What you need

Plasticine or soft clay
A pencil
A rolling pin
Several cylindrical objects

What you do

Work the plasticine until it is soft, then divide it into 2 pieces.
Flatten 1 piece with your fingers, the other with the rolling pin.
- What were the advantages of using your fingers?
- What were the advantages of using the rolling pin?

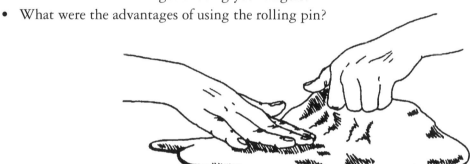

What you do next

Roll up the 2 pieces of plasticine again.
This time, flatten 1 piece with the pencil, the other with the rolling pin.
- What were the advantages of using the pencil in order to roll out the plasticine?
- What were the advantages of using the rolling pin?

Try other round objects in order to roll out the plasticine.
- Which objects work?
- Why do they work?
- How does their size, shape or weight affect their performance?

14 Wheels

Where do we use wheels in the kitchen in order to make work easier?

What you need

An egg beater
A can opener with wheels
A pizza cutter
Cardboard
A pair of scissors for cutting out your model
A round tin
Glue

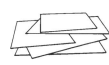

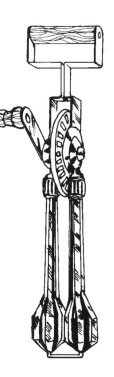

What you do

Look carefully at each tool you have.
Draw each tool, showing any wheels it has.
- How many wheels are there?
- What do the wheels do?
- Are they smooth?
- How do these wheels make work easier?

What you do next

Build a cardboard tool that has wheels that would help you do a household task more easily.

15 Inventing tools

What makes a good tool?

What you need

A pen or pencil for drawing
Paper for drawing and writing
Newspaper
Pieces of cardboard
Adhesive tape
A pair of scissors

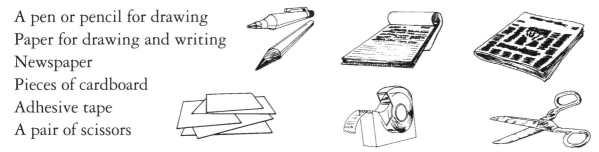

What you do

Draw all the tools you can find in one room in your house.

- How do these tools work?
- What work do they do?
- What do they have in common?
- How do they make work easier?

Group together the tools that work in a similar way.

What you do next

Think of some work you would like to make easier.
Use the newspaper, adhesive tape, cardboard, scissors and anything else you can think of in order to invent a tool that accomplishes this task.
Give your tool a name, and write instructions for its use.

Investigations 16 to 20: Key ideas, and resources needed

Area of focus: The human body

Investigation	Key ideas	Resources needed From home	Resources needed From school
16 Human skeletons	• Bones give structure and strength to bodies. • Bones are of different shapes and sizes. • We can see the places at which ligaments and tendons join to bones. • The ends of some bones are shaped in order to enable joints to move smoothly.	• A pair of scissors	• Split pins • A roll of adhesive tape • Pieces of cardboard • *Master Sheet E: The human skeleton* • A cooked chicken carcass with several types of bones. Boil some chicken carcasses for several hours until the flesh is falling away from the bones. Separate the bones from all the fleshy parts and wash them thoroughly. Dry the bones in a slow oven or in the sun. When storing them, leave them in an open container. • Re-usable adhesive (e.g. Blu-Tack)
17 Lung capacity	• Muscles and the diaphragm contract and relax in order to enable lungs to fill and empty the air we breathe. • Each person's lung capacity can be measured.	• A bucket of water • An empty clear-plastic bottle of at least 2 litre capacity	• A bendable straw • A roll of masking tape • A waterproof marker pen
18 My beating heart	• Heart rate can be measured through observation of the pulse rate. • Heartbeat rate varies according to the amount of exercise being done.	• A watch or clock with a 'second' hand • A pen or pencil for recording	
19 Reflexes	• Messages sent to the central nervous system instigate responses. • With practice, the speed of a person's reaction time can be trained to increase.	• A ruler with a hole drilled in one end • String—length about 3 metres	
20 Getting the picture	• The color of the light entering the eye determines the perception of color. • Colored filters make colors appear to be other than what they are.		• Pieces of cardboard • Red and green cellophane paper • A pair of scissors • A roll of adhesive tape

16 Human skeletons

What are bones like?

What you need

  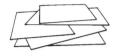

A pen or pencil for drawing
Paper for drawing
Pieces of cardboard
A pair of scissors
Split pins
Adhesive tape
Master Sheet E: The human skeleton
Cooked and dried chicken bones
Re-usable adhesive (e.g. Blu-Tack)

What you do

Look closely at the chicken bones.
Draw them, showing as much detail as you can.
Join them together with the re-usable adhesive to make a part of the chicken, for example a wing.

- What do the bones look like?
- Where might ligaments, cartilage or muscles have been attached?
- Which parts of the bones fit together well?
- How would these parts have moved?

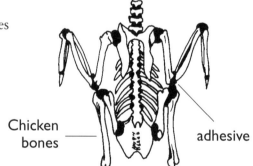

Chicken bones — adhesive

What you do next

Look at *Master Sheet E: The human skeleton*.
Using the cardboard, scissors, split pins and adhesive tape, build a skeleton.

Look at *Master Sheet E: The human skeleton*.

- What are the bones' anatomical names?
- What are the bones' common names?
- Which parts of the bones can you feel through your skin and muscles?
- What types of joints do you have?

17 Lung capacity

How much air do I take in each time I breathe?

What you need

A bucket of water
An empty clear-plastic bottle of at least 2 litre capacity
A bendable straw
A waterproof marker pen

What you do

Predict and record how much air you exhale each time you breathe out.
Completely fill the bottle with water and screw the lid on.
Half-fill the bucket with water.

Tip the bottle upside down in the bucket so the bottle neck is underwater. Take the lid off. Put one end of the bendable straw into the bottle neck.

Take a deep breath and breathe as much air as you can into the bottle. Put the lid back on the bottle.

Measure how much water has been pushed out of the bottle by the air you breathed into it.

Try this several times in order to find the average.
Each time, make sure the bottle is filled with water before you start.
- How great is your lung capacity?

What you do next

Use reference books in order to find out what lungs look like and how they work, then record the information.

Test other people's lung capacity, using a clean straw for each person.
- How does the size of the person's chest relate to his/her lung capacity?

18 My beating heart

What happens to my heart when my body works hard?

What you need

A clock or watch with a 'second' hand
A pen or pencil for recording
Writing paper

What you do

Find your pulse.
You can measure your heart rate by counting your pulse. Count each beat for 30 seconds, then double the time to 60 seconds in order to find out how many beats there are per minute.

What you do next

Lie still for 3 minutes.
Measure your pulse for 1 minute.

Stand up and walk around for 3 minutes.
Measure your pulse for 1 minute.

Run as fast as you can for 3 minutes.
Measure your pulse again.

Sit quietly for 3 minutes.
Measure your pulse again.

- When does your heart beat the slowest?
- When does it beat the fastest?
- What happens to your heart rate after you have stopped exercising?
- What happens to your heart rate as you increase your activity?
- How is your heart affected by your activity?

Use reference books in order to find out about how your heart works and what it looks like.

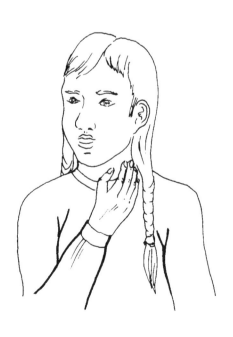

19 Reactions

How fast are your reactions?

What you need

A ruler with a hole drilled in one end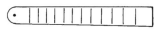
String

What you do

1. Tie the ruler to a fixed point so it hangs freely.
2. Have someone hold the ruler while your hand is not grasping but is level with the 10 centimetre point.
3. As the ruler is let go, try to grab it as quickly as you can.
4. Measure the distance the ruler fell before you grabbed it. This is a measure of your reflex time.
5. Repeat this 10 times and average out your reaction times.
   - How quickly were you able to react?
   - What was the best thing to look at in order to improve your reaction time?

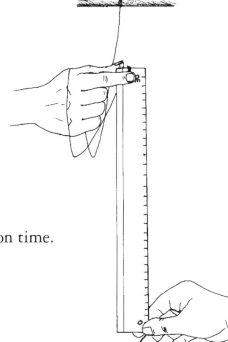

What you do next

Test and record your family members' reaction times.
- Who was the quickest?

Find out whether practice improves reaction time. Practise 10 times each day for one week, then measure your reaction time again.
- When were you able to react the quickest?
- Does practice improve your reaction time?

Use reference books in order to find out more about reactions and reflexes and how they work.

Devise other reaction-time testers.

20 Getting the picture

How do you see colors?

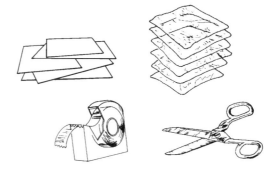

What you need
Pieces of cardboard
Red and green cellophane paper
Adhesive tape
A pair of scissors

What you do
Using the red cellophane paper, make a pair of glasses.

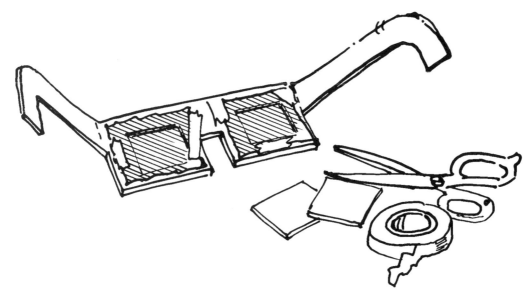

Look at various things through the glasses.
- What color do red things seem to be when you have the glasses on?
- What color do other colors seem to be?
- Which colors seem to be black?
- How does wearing the glasses change the things you see?

What you do next
Using the green cellophane paper, make another pair of glasses.
- What color do various things seem to be when you have these glasses on?

Using red cellophane paper for one eye and green cellophane paper for the other eye, make another pair of glasses.
- How does this affect your sight?

Investigations 21 to 25: Key ideas, and resources needed

Area of focus: **The sky**

Investigation	Key ideas	Resources needed From home	From school
21 The moon	• The moon has a regular cycle of movement around the earth. • The appearance of the moon and the time it is visible in the sky change in a regular and predictable way.	• A pen or pencil for recording • Writing paper	
22 The stars	• We can identify and name stellar constellations. • We can see and record apparent star movements.	• A pen or pencil for recording • Writing paper	• *Master Sheet D: The constellations of the northern sky*
23 The sun	• The sun seems to rise in the east and set in the west in all countries.	• A pen or pencil for recording • Writing paper	• A compass
24 Clouds	• Clouds have characteristics we can see and record. • We can classify clouds according to their color, shape and density.	• A pen or pencil for recording • Writing paper	• *Master Sheet F: Cloud formations*
25 Day and night	• The times of the sun's rising and setting change. • We can predict the times, and they follow a regular pattern throughout a year.	• An alarm clock • A pen or pencil for recording • Writing paper	

21 The moon

When can you see the moon in the sky?

What you need

A pen or pencil for recording
Writing paper

What you do

Every day for one week, look for the moon.
Note the time you see it and the shape it is.
- What time does the moon rise each day?
- What shape is it?
- How does its appearance change each day?

What you do next

Use the chart you have made in order to predict what the moon will look like for the following week. Include the time you think it will rise.

Check your predictions.
- How accurate were they?
- Does the moon change from

or

Use newspapers and calendars in order to find out about the moon's phases.

22 The stars

Which stars do you see in the sky?

What you need

Master Sheet D: The constellations of the northern sky
A pen or pencil for recording
Writing paper

What you do

At sunset, go outside and watch the stars as they appear.
Record any patterns or interesting combinations you see.

- Which stars seem to be close together?
- Which do you see first at night?
- Which are the brightest?

What you do next

Look at *Master Sheet D: The constellations of the northern sky*.
Find some of the stars shown on it.
- What is the name of the star you saw first?
- What are the names of the constellations you drew?

Use reference books in order to find out more about stars.
- How can you tell a star from a planet?

23 The sun

Where does the sun rise and set?

What you need

A compass
A pen or pencil for recording
Writing paper

What you do

Each day, stand in the same place in order to find out and record the direction you have to look towards to see the sun rise and set.

Warning: Do not look directly at the sun—you will damage your eyes.

- What time did the sun rise each day?
- What direction did you have to look towards?
- How did the sky change as the sun rose?
- How did the sky change as the sun set?

What you do next

Find out about why the sun seems to move around the earth.
Build a model or draw a picture that explains this.

24 Clouds

What types of clouds are there?

What you need

A pen or pencil for recording
Writing paper
Master Sheet F: Cloud formations

What you do

Each day, draw pictures of the clouds you see in the sky.

- What color are the clouds?
- How much of the sky do they cover?
- Are they moving?
- How fast are they moving?
- What type of weather do you get when the clouds look this way?
- What type of weather do you get the day after?

What you do next

Make a weather chart showing what the cloud cover has been.

Use reference books in order to find out what makes clouds form.

25 Day and night

Does the sun seem to rise and set at the same time each day?

What you need

An alarm clock
A pen or pencil for recording
Writing paper

What you do

Predict what time you will be able to first see the sun and last see the sun each day for 1 week. Record your predictions.

Each morning, set your alarm clock to wake you just before you think you will be able to first see the sun. Record the actual time you could see the sun each day.
- How closely did you predict your first view of the sun?
- What time did you first see the sun each day?
- Was the time the same each day?
- How much did the time differ from the start of the week to the end of the week?

	MON	TUE	WED	THUR	FRI	SAT
SUN RISE	6.47	6.48				
SUN SET	7.36	7.37				

What you do next

Predict what time you think you will last see the sun each day. Record your predictions.

Each evening, record the actual time you last saw the sun.
- How closely did you predict your last view of the sun?
- What time did you last see the sun each day?
- Was the time the same each day?
- How much did the time differ from the start of the week to the end of the week?

Check your records against the ones published in the newspaper.
- Are your records the same as those in the newspaper?
- What could account for the difference?

Investigations 26 to 30: Key ideas, and resources needed

Area of focus: Playground equipment and toys

Investigation	Key ideas	Resources needed — From home	Resources needed — From school
26 Swings	• Swings behave in regular and predictable ways. • A swing's height of release does not affect the number of swings in 1 minute. • The swing's mass does not affect its rate of swing.	• A clock or watch with a 'second' hand • A pen or pencil for recording • Writing paper • Another person	
27 Sliding down	• An object will take the same time to slide down the same slope each time it is released. • We can use slopes in order to move objects.	• A tennis ball • A slide (slippery dip) • Another person • A clock or watch with a 'second' hand	
28 Bouncing balls	• A dropped ball will rebound to a height less than that it was dropped from. • As the ball strikes a hard surface it is compressed. • As the ball returns to its shape it pushes against the surface and rises.	• Several balls of different sizes, for example a tennis ball, a basketball and a small rubber ball	• A tape measure • Re-usable adhesive (e.g. Blu-Tack)
29 Balls and surfaces	• The nature of the surface a ball is dropped on determines the height the ball will bounce. • Some materials absorb more energy than others. • The material a ball is made of will determine the height of its bounce.	• A tennis ball	• A ruler of length 1 metre or a stick marked 1 metre
30 Rolling down hills	• We can use slope in order to move objects. • As a slope's gradient increases, the distance travelled by a ball rolling down it also increases.	• A tennis ball • A book • Several building blocks all of the same size	• A tape measure

26 Swings

How many times will a swing swing in 1 minute?

What you need

A clock or watch with a 'second' hand
A pen or pencil for recording
Writing paper
2 other people of different weights

What you do

Hold the swing up level with your nose.
Let it go and step out of the way.
Count how many times it swings in 1 minute—one swing is down and back again.

Try it again, starting from holding the swing level with your waist.
- How many swings does it make in 1 minute?
- How did the height of the swing's starting position affect the number of swings?

What you do next

Repeat the same experiment, but have one of the people on the swing.
- How many swings are there in 1 minute?

Try it again, but have the other person on the swing.
- How many swings are there in 1 minute?
- How does the weight of the person on the swing affect the number of swings?

27 Sliding down

How does slope aid movement?

What you need

2 balls of different sizes, for example a tennis ball and a basketball
A slide (slippery dip)
Another person
A clock or watch with a 'second' hand

What you do

Hold one ball at the top of the slide and let it go.
Time how long it takes the ball to roll down the slide.

Try this several times in order to find the average time.
Now try it with a different ball.
- How long does it take this ball to get to the bottom?
- What things affect a ball's speed when it is rolling down a slope?

What you do next

Find out how long it takes you to slide to the bottom of the slide.
Compare your times with those of the other person.
- What fabric were you both sitting on?
- Who was the fastest?
- Who is the heaviest?
- What things affect speed down the slope?

28 Bouncing balls

How high will a ball bounce?

What you need

Several balls of different sizes, for example a tennis ball, a basketball and a small rubber ball
A tape measure
Re-usable adhesive (e.g. Blu-Tack)

What you do

Attach the tape measure to an outside wall that has concrete next to it.
Choose one of the balls you have.
Drop this ball from a height of 1 metre and watch how high it bounces on its first bounce.
Try this several times in order to be sure.

Try it again, this time dropping the same ball from a new height of, say, 1.5 metres on to the same surface.
- What happens to the bounce height now?
- How does the height from which a ball is dropped affect its bounce?
- How high does this ball bounce after it is dropped?

What you do next

Repeat with the other balls.
- How high does each ball bounce?
- Which ball bounces best?

29 Balls and surfaces

Which surface is best for ball games?

What you need

A tennis ball
A ruler of length 1 metre or a stick marked 1 metre

What you do

Hold the ball at a height of exactly 1 metre from the ground.
Drop it next to the ruler and watch to see how high it bounces.

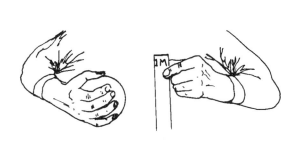

Try this several times in order to be sure.

What you do next

Try this on as many other surfaces as you can,
for example sand, gravel, concrete, grass and carpet.
- How high did the ball bounce on each surface?
- How did the surface affect the bounce's height?
- Which surface is best for bouncing balls on?

30 Rolling down hills

How far will a ball roll?

What you need

A tennis ball
A book
Several building blocks all of the same size
A tape measure

What you do

In an open flat area, make a slope for the tennis ball.
Let the tennis ball go at the top of the slope and measure how far it rolls before it completely stops.

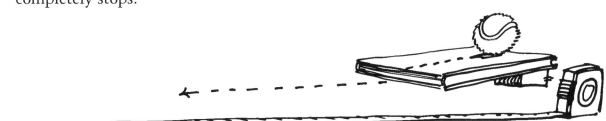

Do this several times in order to be sure.
- How far does the tennis ball roll?

What you do next

Raise the book up by one more block and find out how far the ball rolls now.
Try this with various slopes.
- How far does the ball roll each time?
- How does increasing the slope affect the distance the ball rolls?

Investigations 31 to 35: Key ideas, and resources needed

Area of focus: **Backyards**

Investigation	Key ideas	Resources needed From home	From school
31 The living things in your backyard	• We can classify all things as living, once living or non-living. • Living things have a relationship to their habitats. • We can see and record characteristics of all things.	• A pen or pencil for recording • Writing paper • String • Satay sticks with their points removed	• A magnifying glass
32 Leaves	• Many plants have leaves, and leaves have a diverse range of shapes, colours and textures.	• A pen or pencil for recording • Writing paper	• A magnifying glass • String—about 4 metres • Satay sticks with their points removed
33 Snails on the move	• A snail's habitat provides the snail with shelter, moisture and food. • Snails often shelter in the same place each day. • We can record the movement of snails.	• Nail polish • A map of your backyard • A flashlight • 2 marker pens in different colours	• A magnifying glass
34 Animals in the backyard	• A wide range of animals live in the backyard habitat. • The number of each animal living in a habitat depends on the richness of the food source, availability of shelter, climate, and presence of predators.	• A pen or pencil for recording • Writing paper	• A magnifying glass
35 Soils are soils	• Soil is made up of a variety of organic and non-organic material. • Soil's composition changes as we dig further below the surface. • Soil is made up of small and large pieces.	• A shovel • A bucket • A strainer • An empty yoghurt container	• A magnifying glass

31 The living things in your backyard

What lives in your backyard?

What you need

A magnifying glass
A pen or pencil for recording
Writing paper
String—about 4 metres
Satay sticks with their points removed

What you do

Choose four places in your backyard.
Measure 1 square metre for your observation area.

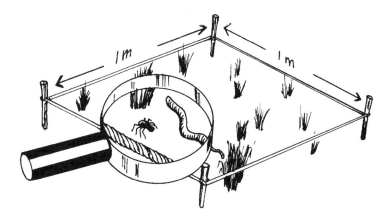

List all the things you see there.
- Which living things did you see?
- Which things were non-living?
- Which things were alive once but not now?
- Which things did you find in one study area but not in the other study areas?
- Which things did you find in all 4 study areas?

What you do now

From each category, choose 2 things you saw.
Using the magnifying glass, look at them more closely.
Draw them, showing as much detail as you can.

32 Leaves

What shape can leaves be?

What you need

A magnifying glass
Pencils for doing rubbings
Writing paper

What you do

Collect a leaf from every type of plant in your yard.
Group your leaves according to your own criteria.

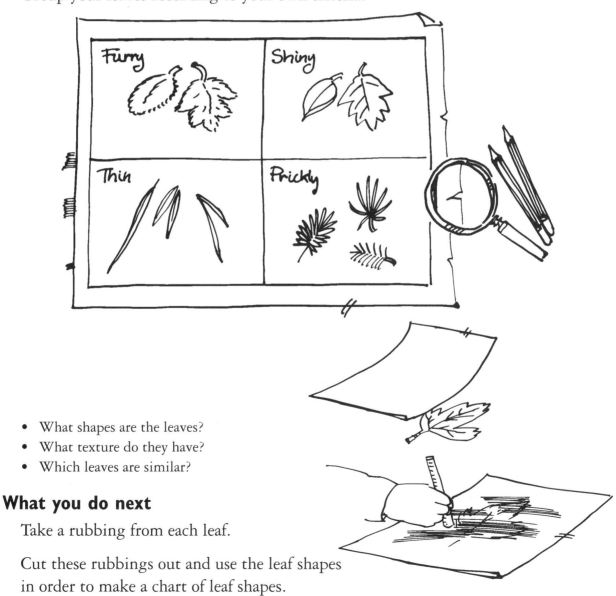

- What shapes are the leaves?
- What texture do they have?
- Which leaves are similar?

What you do next

Take a rubbing from each leaf.

Cut these rubbings out and use the leaf shapes in order to make a chart of leaf shapes.

Page 59 of *Take-Home Science*, © Jenny Feely, 1994. This page may be reproduced for classroom use.

33 Snails on the move

Where do snails go?

What you need

Nail polish
A map of your backyard
A flashlight
2 marker pens of different colours

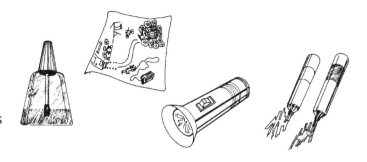

What you do

During the daytime, look for snails in your backyard.

On your map, use one of the marker pens to mark the place you found the snails, as well as how many snails there were.

Carefully put a small amount of nail polish on the snails' shells.

- How many snails did you find?
- Where were they during the day?
- What were they doing?

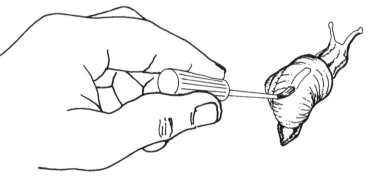

What you do next

At night time, revisit each place at which you recorded snails were during the day. Check to see whether the snails are still there.

Look for the marked snails in other places in the yard.

On your map, using the other marker pen, record the places you find the marked snails.

- How many snails were in the same place?
- Where else did you find the marked snails?
- What were they doing?

Look for the marked snails again in the morning.

- Where are the snails now?
- What are they doing?
- How many of the marked snails could you find?

34 Animals in the backyard

Which animals live in your backyard?

What you need

A magnifying glass
A pen or pencil for recording
Writing paper

What you do

Find out which animals live in your backyard.
Warning: Do not handle these animals—some of them can be dangerous.

Record the animals you see, and try to work out how many of them live in your backyard.

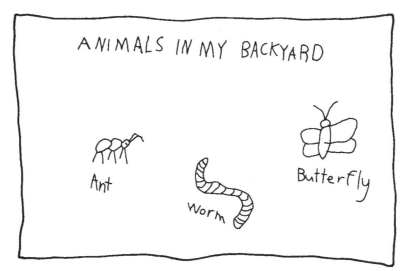

What you do next

Choose one of these animals to study more closely.
Draw the animal, showing as much detail as you can.
Show the animal's habitat.

Use reference books in order to find out as much as you can about the animal.

35 Soils are soils

What is in the dirt in the backyard?

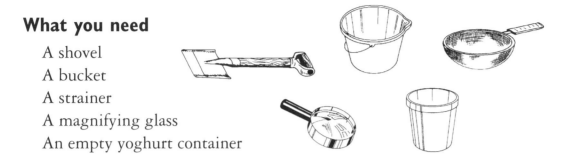

What you need

- A shovel
- A bucket
- A strainer
- A magnifying glass
- An empty yoghurt container

What you do

Choose a place in your backyard out of which you are able to dig a small amount of dirt.

From the top layer of the soil, take enough dirt to fill the yoghurt container.

Sift the dirt into the bucket.

Look carefully at what you have found.
- What is in the bucket?
- What has been left in the strainer?

Record your findings.

What you do now

Dig down a little further and take some more soil from further down.

Sift this soil.
- What do you find in this sample?
- How is it the same as the first sample?
- How is it different?

Dig down further and repeat your test.
- How does the soil change as you dig deeper?

You may like to try this in other areas of the yard.

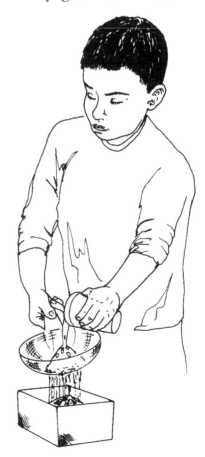

Investigations 36 to 40: Key ideas, and resources needed

Area of focus: **Plants**

Investigation	Key ideas	Resources needed — From home	Resources needed — From school
36 Growing beans	• Seeds contain enough energy to start growing. • Bean seeds send out roots that move downwards and shoots that move upwards.	• A glass jar • A bucket • Tap water • A shovel • A pen or pencil for recording • Writing paper	• A roll of paper towel • A magnifying glass • 4 bean seeds—keep them in empty film cases for taking home. Having 1 packet of seeds should enable all the children in your class to complete the investigation.
37 Growing potatoes	• Potatoes are tubers. • Potatoes contain enough nutrients to make a new plant start growing. • Potatoes reproduce by forming new tubers.	• A potato with 'eyes' • A pen or pencil for recording • Writing paper • A shovel	• A magnifying glass
38 Growing plants from cuttings	• Some plants can be grown from cuttings and water. • Some plants will form roots when parts of them are exposed to water. • We can plant the cuttings in order to grow new plants.	• A glass jar • A pair of scissors • Tap water • A pen or pencil for recording • Writing paper	• Some leaves from plants such as geraniums, daisies and African violets • Plastic wrap • A rubber band • A satay stick
39 Good growing conditions	• Plants need water, nutrients and some light in order to grow in a healthy way.	• 6 yoghurt pots • A sample of soil • Several stones • Tap water • A pen or pencil for recording • Writing paper • A pair of scissors	• Bean seeds—keep about 12 beans in an empty film case for taking home. Having 10 packets of beans should enable all the children in your class to complete the investigation.
40 Plant parts	• Plants have parts we can identify. • Flowers can be of various shapes and colours. • Some plants have flowers that are their reproductive parts.	• A pen or pencil for recording • Writing paper • 6 flowering plants for drawing their parts	• A magnifying glass • *Master Sheet G: The parts of a flower*

36 Growing beans

How do bean seeds grow?

What you need

A glass jar
A bucket
Tap water
4 bean seeds
Some paper towel
A magnifying glass
A pen or pencil for recording
Writing paper
A shovel

What you do

1 Make a cylinder of paper towel and place it in the jar.
2 Carefully place the bean seeds between the jar and the paper towel.
3 Carefully stuff some more paper towel into the jar, making sure you do not move the paper cylinder or the beans.
4 Put some water into the jar— enough for the paper towel to soak up and become completely wet. Add a little water first, then wait and watch in order to see how much you need.

5 Put the jar on a window ledge or another place where it will get good light.

What you do next

Each day, check your paper towel is still damp. Add more water if needed.
Look carefully at the bean seeds.
- What do you see happening?
- Have the beans changed?
- How have they changed?

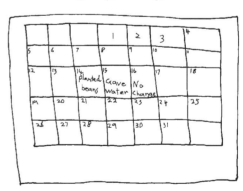

Keep daily records of what you see happening.
After about 3 weeks you will be able to plant your beans in the garden.
You might like to continue your observations of the bean seeds' changes after planting.

37 Growing potatoes

How do potatoes grow?

What you need

A potato with 'eyes'
A magnifying glass
A pen or pencil for recording
Writing paper
A shovel

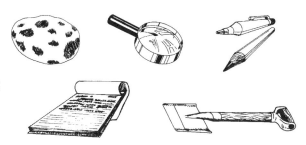

What you do

Put your potato on a window ledge.
Every couple of days, look
at the potato and record any changes you see.

- What happens to the potato?
- How does it change?

What you do next

Once your potato has sprouted leaves, plant it in the garden.
Observe what happens as time passes.
After the plant dies, dig up the roots.
- What do you find?

38 Growing plants from cuttings

How can plants grow from small parts?

What you need

- A glass jar
- Some leaves from plants, for example geraniums, daisies and African violets
- A pair of scissors
- Plastic wrap
- A rubber band
- A satay stick for making holes in the plastic wrap
- Tap water
- A pen or pencil for recording
- Writing paper

What you do

1. Carefully cut several leaves from the plant you have chosen.
2. Using the scissors, cut the stem diagonally so about 4 centimetres of it are left attached to the leaf.
3. Fill the jar with water, then stretch the plastic wrap over the mouth of the jar. Place the rubber band around the jar in order to hold the wrap in place.
4. Using the satay stick, carefully make small holes in the plastic wrap, then place the leaf cuttings in so the ends are under the water.

What you do next

Leave the cuttings for a few days.

Look at them regularly and record any changes that occur.

- What happens to the leaves?

If any of the cuttings develop good roots, you can plant them in some potting mix.

39 Good growing conditions

What do plants need in order to grow?

What you need

6 empty yoghurt containers
A pair of scissors for poking holes in the containers
A sample of soil
Several stones
About 12 bean seeds
An empty film case for taking the bean seeds home
Tap water
A pen or pencil for recording
Writing paper

What you do

Carefully poke 4 small holes in the bottom of each yoghurt container.
Fill 5 of the containers with the soil and 1 with the stones.
Plant several beans in each container.

What you do next

Every few days, check your containers.
Record what you see.
- What happens to the beans that get no water?
- What happens to the beans that get no light?
- What happens to the beans that have no soil?

- What do beans need in order to grow well?

40 Plant parts

What parts do flowering plants have?

What you need

6 types of flowering plants
A magnifying glass
Master Sheet G: The parts of a flower
A pen or pencil for recording
Paper for drawing

What you do

Choose 3 types of flowering plants in your yard.
Draw them.
Using *Master Sheet G: The parts of a flower*, label the parts. Count them.
- How are these plants the same?
- How are they different?

What you do next

Choose 3 flowers from different plants.
Carefully pull them apart and look at the parts.
Draw a diagram of each flower, showing its parts.
- How are the parts the same?
- How are they different?

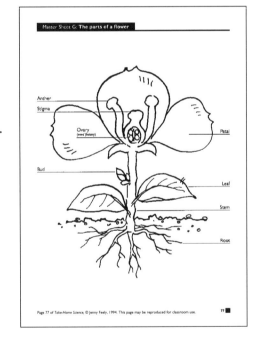

APPENDIX

Master sheets

Master Sheet A: Log-book record

Date	Investigation number	Investigation title	Comments

Master Sheet B: Possible ways to record your findings in your log book

1. Draw what you find out, or write about it.

2. Write a report that includes the following parts.
   - Purpose—what you were trying to find out
   - Prediction—what you thought would happen
   - Procedure—what you did
   - Results—what you found happened
   - Conclusion—what you think your results mean

3. Write answers to the questions on the activity cards.

4. Use tables.

5. Use diagrams.

6. Write a poem or a story about what you have learnt.

7. Use graphs.

8. Write a letter.

9. Keep a diary.

10. Draw a comic strip showing everything you did and saw.

Master Sheet C: 'Activities I have completed'

- Mark off each activity when you have finished it.

My name: ....

1	2	3	4	5
6	7	8	9	10
11	12	13	14	15
16	17	18	19	20
21	22	23	24	25
26	27	28	29	30
31	32	33	34	35
36	37	38	39	40

Master Sheet D: The constellations of the northern sky

Master Sheet E: **The human skeleton**

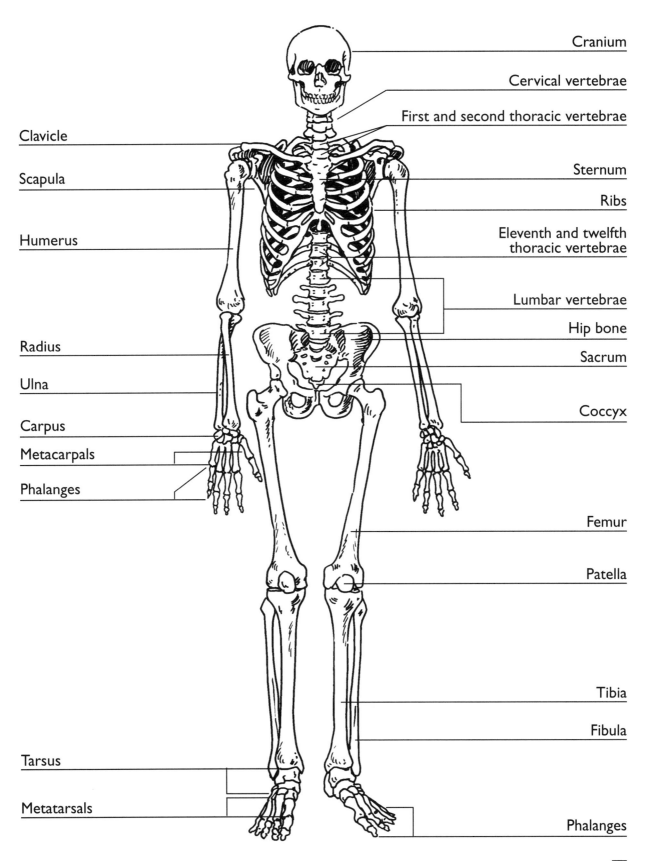

Master Sheet F: Cloud formations

Master Sheet G: The parts of a flower

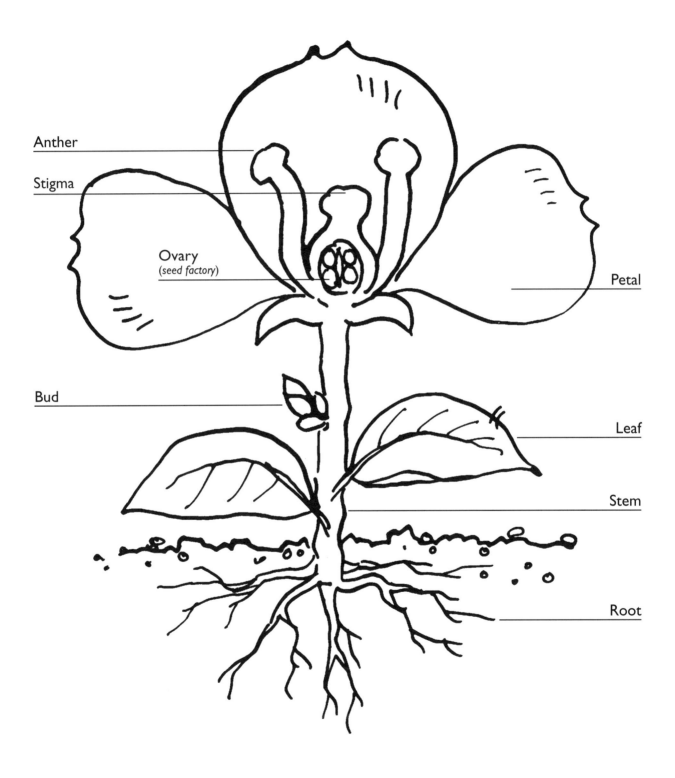

Index

Animals 58, 60, 61

Balls 54, 55

Clouds 49

Day and night 50

Elasticity 54
Energy 52, 53, 56

Food 31

Gravity 53, 54

Habitat 58, 60
Heart 42
Human body 42, 44

Levers 35
Light 44
Living things 58, 59, 60, 61, 64, 65, 66, 67, 68
Lungs 41

Materials 28, 29, 30, 31, 32
Moon 46

Packages 28, 29, 30, 31, 32
Plants 58, 59, 64, 65, 66, 67, 68
Play equipment 52, 54, 55, 56

Reactions 43
Rollers 36

Shapes 22, 23, 24, 25
Skeleton 40
Soil 62
Stars 47
Structures 22, 23, 24, 25, 26
Sun 48

Temperature 30
Tools 34, 35, 36, 37, 38

Wedges 34
Wheels 37

Other books from Heinemann

The Home-School Connection
Guidelines for working with parents
Jacqueline McGilp and Maureen Michaels
The benefits of stronger ties between school and home are backed by research and acknowledged by both parents and teachers. But establishing effective connections between home and school is not always easy. *The Home-School Connection* takes on the challenge of getting the home and school to work together, to listen to each other and to use resources well so that effective learning for students is a reality. The book raises issues and establishes guidelines for involving the home and the wider community in educating children.

The *Home-School Connection* caters for a wide range of involvement, from small steps for the individual teacher to whole school strategies and provides practical ideas and activities.
ISBN 0 435 08820 3   illustrated   96pp

Thinking for Themselves
Developing strategies for reflective learning
Jeni Wilson and Lesley Wing Jan
By encouraging children to think about their learning and to become aware of and control their thinking processes, teachers can help children to become active, responsible learners - learners who can make their own decisions, choose appropriate strategies, assess their own work and set their own goals. In a book filled with activities for the development of skills and strategies within a range of existing programs, well-known authors and educators Jeni Wilson and Lesley Wing Jan provide starting points and ideas to help the implementation of reflective teaching and learning programs.
ISBN 0 435 08805 X   illustrated   156pp

I Teach
A guide to inspiring classroom leadership
Joan Dalton and Julie Boyd
Specific and practical insights into the 'what' and 'how' of effective learning and teaching presented succinctly and visually.

Contents: Identify your goals • Walk the leader's walk • Build relationships with others • Create a community of learners • Empower growth in others • Work on self-growth - identify personal strengths, highlight areas for self-improvement and plan for balanced leadership.
ISBN 0 435 08782 7   illustrated   144pp

Math Makes Sense
Teaching and learning in context
Rachel Griffiths and Margaret Clyne
Maths Makes Sense shows ways of teaching mathematics in contexts that make sense to children and that therefore help children make sense of mathematics. This is a very 'hands on' book for teachers who want to 'do' mathematics rather than read about the subject.

Teachers are given examples of the principles and strategies at work in the classroom. These examples are accompanied by activities to use on the spot, so that teachers can see for themselves how exciting and effective teaching mathematics in context can be.
ISBN 0 435 08362 7   illustrated   144pp

Responsive Evaluation
Making valid judgments about student literacy
Brian Cambourne and Jan Turnbill, eds
Changes in teaching practice caused by new understandings of how children learn have made many traditional methods of evaluation obsolete. Demands on teachers and educators to demonstrate accountability have put pressure on these same educators to devise new methods of assessment that demonstrate accountability and are appropriate to current teaching methods. Procedures must be established that lead to optimum learning, reflect wholistic thinking, enrich classroom teaching and are seen to be rigorous, scientific and valid.

Jan Turnbill and Brian Cambourne have worked with teachers, principals, academics, parents and students to establish assessment procedures that fit these guidelines. They have all contributed to *Responsive Evaluation* and report on how they put theory into practice.
ISBN 0 435 08829 7   illustrated   144pp

Raps & Rhymes in Maths
Compiled by Ann and Johnny Baker
A collection of traditional and modern rhymes, riddles and stories with mathematical themes, *Raps & Rhymes in Maths* can be used to provide a welcome break from more formal activities or can form the introduction or conclusion of a maths lesson. The raps, rhymes and stories provide openings for mathematical investigations and, most importantly, provide a source of enjoyment.
ISBN 0 435 08325 2   illustrated   90pp

Counting on a Small Planet
Activities for environmental mathematics
Ann and Johnny Baker
Counting on a Small Planet uses mathematics to explore and explain how our actions affect the world we live in and develops children's awareness of the responsibility we each have for looking after our environment. The suggested investigations provide ideas for further action, and the Maths Fact Files at the end of each topic relate the local to the global situation. • All That Rubbish • Be Quiet • Water Use and Abuse • Not a Drop to Spare • When the Wind Changed • Design a House
ISBN 0 435 08327 9   illustrated   98pp

Maths in the Mind
Ann and Johnny Baker
Maths in the Mind looks at the acquisition of basic number facts and focuses on the development of mental skills and strategies within the context of broader activities. Twenty fully developed activities provide the children with the practice needed to develop, access and recall number facts speedily.
ISBN 0 435 08316 3   illustrated   120pp

Mathematics in Process
Ann and Johnny Baker
Mathematics in Process investigates the purposes and conditions of learning and doing mathematics.
   Part One looks at the child's experience and how children get involved, how young mathematicians work and how children communicate and learn from reflection.
   Part Two presents ideas on devising a curriculum and how to set up the classroom, and features a complete section of activities to try immediately with your class.
ISBN 0 435 08306 6   illustrated   176pp

Maths in Context
Deidre Edwards
Maths in Context shows how to integrate mathematics with the wider curriculum areas by using a central theme. This approach increases children's motivation, caters for individual differences and increases confidence in mathematical ability. Mathematics is seen as part of 'real life'.
   A large section of the book presents ideas for activities based on Dragons, Our Environment, The Zoo, Party Time, Traffic, Christmas, Show and Tell and The Faraway Tree.
ISBN 0 435 08308 2   illustrated   152pp